Teaching Multiplication and Division of a
2-DIGIT NUMBERS by 1-DIGIT NUMBERS

Frank Gardella

Cover image © Shutterstock.com

www.innovativeinkpublishing.com
Send all inquiries to:
4050 Westmark Drive
Dubuque, IA 52004-1840

Copyright © 2024 Francis Gardella

Print ISBN: 979-8-3851-2710-8
Ebook ISBN: 979-8-3851-2711-5

Published in the United States of America

Contents

Teaching Multiplication of a 2-Digit by a 1-Digit Whole Number Using Base-10 Materials 1

Multiplication: Trading Tens for Ones

Teaching Division of a 2-Digit by a 1-Digit Whole Number Using Base-10 Materials

Division: Trading Tens for Ones

Introduction

The use of physical models is extremely helpful in assisting students at all levels understand the mathematics that they are studying. The goal of instruction is to have the students use this understanding to work with mathematics at the symbolic level.

This book will assist you as you guide your students through the process of using what they learn from modeling mathematics with Base Ten materials to the recording of this work using the standard mathematical symbols and algorithms for the Whole Number Operations of Multiplication and Division.

What Are Base 10 Materials?

Base 10 Materials are models that physically can represent the values in our place value system.

FOUR BASIC MODELS

There are four types of pieces in a set of **Base 10 Materials** when dealing with Whole Numbers:

i.   A cube which represents a value of '1'. This is called a 'one.'

ii.   A bar of 10 cubes which represents a value of '10.' This is called a 'ten.'

iii.   A flat of 100 cubes which represents the value of '100.' This is called a 'hundred.'

iv.   A large cube or block containing 1000 cubes which represents the value of '1000.' This is called a 'thousand.'

Since the 'thousand' is in 3-dimensions, we cannot create a geometric representation of any place value higher than 'thousands.'

NOTE: However, there are materials, such as Chip Trading developed by Patricia Davidson which uses colored chips to represent place values. Using these materials, the place values can be extended beyond thousands.

REPRESENTING QUANTITIES

To represent the number 27, you can use 27 cubes (show picture of 27 cubes)

or you can use 2-tens and 7-ones.

In using these materials, we can allow students to show numbers using any combination of the pieces. However, when we begin to link the pieces to the mathematical symbols, we encourage students to use as few pieces as possible. As you will see in the following pages about the **Base 10 Materials** being used to model the operations of multi-digit numbers, it is best to have students use the **least number of pieces**. So in the case of 128, the students would use 1-hundred, 2-tens and 8-one. This idea can be called **The Rule of Least Pieces** (Gardella, 2008) which stems for the work of Matthew Scaffa and Janet Castellano who were the Mathematics Supervisory Team for the New York City Public Schools on Staten Island in the 1970's through the 1990's.

The idea of "Least Pieces" gives a physical reason for the value in any place in our system never being more than 9.

Reference:

Gardella, F. (1996). *Least pieces: A Manipulative Necessity.* The New Jersey Mathematics Teacher, 54 (1), 16-19

Some History of Physical Models in Mathematics

Materials to physically model ideas of number have been used since people began using stones to represent quantities before symbolic numerals were developed. While the ideas of learner involvement stand prominently in education through the work of many leaders such as Jean Piaget, Lev Vygotsky and Jerome Bruner, the modern classroom use of physical models in learning mathematics can be traced to three people who worked in the same time period in the middle of the 20th Century. These people were Zoltan Dienes, a mathematician, Catherine Stern, a Montessori-trained educator and George Cuisenaire, a musician and educator.

Zoltan Dienes (1916-2014) was born in Hungary but studied and conducted his early teaching in England. While teaching mathematics at the University of Leicester, he also attended courses in the education department and earned a Diploma in Education there in 1953. This led him to study Supplementary Psychology at the University of London. As he moved along in his career, he felt more and more that the mathematics that he saw as beautiful and exciting was, through being taught incorrectly, seen by most as scary and boring. It was the connection between his background in mathematics and psychology that had him begin to investigate how this connection could assist in the learning of mathematics.

Through the influence of Emile Borel, Dienes came to recognize how the personality of children impacted on their conceptualization of mathematics and that constructive thinking on the part of children was much more of a characteristic of children than the analytical thinking usually associated with learning mathematics. With this, he proposed the idea that teaching materials used should be based on the mathematical subject addressed rather than students' developmental stages based on their ages. This was in concert with his idea that by using manipulative materials, games and stories, children can understand more complicated mathematics at a younger age than had previously been thought.

Through his work as a member of the International Study Group of Mathematics Learning, he led research in the learning of mathematics by children between six and twelve years old.

Throughout his life, Dienes seemed to balance his work in mathematics with the way children learn mathematics and worked with elementary mathematics teachers wherever he was located.

This is why there are those who refer to Base-10 materials as Dienes Bars.

Catherine Stern (1894-1973) was a teacher and administrator in Breslau Germany in the late 1920's. She began to view the teaching of mathematics in a more structural fashion. While the Montessori Mathematics Materials used beads to make strings (such as 4 beads and 5 beads yielded a string of 9 beads), it made counting the individual beads the primary method for obtaining an answer of 9. Stern saw this differently. She glued 1-cm cubes together in a row and created linear number blocks in different colors. So, a block of 4 cubes and a block of 5 cubes was equal to a block of 9 cubes.

Emigrating to the United States in 1938, she introduced her materials at the Winward School in White Plains, NY. A major difference in her work here was that she used 1-inch cubes for the number bars as opposed to the 1-cm cubes used in Europe. As a result, in the United States today, these inch-bars are recognized as Stern Blocks.

Through her work with the psychologist Max Wertheimer, she realized that rote learning did not give learners insight into relationships. She saw that the visualization of the structural characteristics of the concept involved provided the true relationships of concepts that was needed by the learner.

Together with her daughter, Toni Gould and a student from Bank Street College, Margaret Bassett, the work of Catherine Stern was expanded to create published works that then became part of her Structural Arithmetic Program. A noted professional work was the book, Children Discover Arithmetic, co-authored with Margaret Bassett Stern. This writing sets the foundation for her work in helping children learn mathematics.

Her work addresses number ideas for simple numbers as well as the use of multiple materials in dealing with operations with multi-digit numbers.

More information relative to her program, called Stern Math, can be found at http://www.sternmath.com

Georges Cuisenaire (1891 – 1975) studied the violin and obtained both his license to teach school as well as music. He spent his professional life teaching in his native Belgium in Lower Town (Brussels) and then as Director of Primary Education. As a teacher he had a belief that, because of a child's affinity to color, the frequency ratios that exist in sounds can be developed using colors. In 1945 he began to produce sets of colored cardboard strips which he found very useful for learning arithmetic. Like Catherine Stern, he used definitive sized centimeter lengths to determine a number. In 1952, he produced his first and most well known work, "Les Nombres en Couleurs" (Numbers in Colour). He then met the psychologist, Caleb Gattegno who helped make the use of rods in teaching mathematics more widely known. Gattegno also saw that the rods allowed children to investigate aspects of mathematics on their own. Together, they published an account of the rods in English which led to the rods coming to the United States. In his work, Cuisenaire was also supported by Georges Papy who along with his wife, Frederique, was influential in the development of the Comprehensive School Mathematics Project by Fred Kaufman in the 1960's. This program is archived at the University of Buffalo.

Today, Cuisenaire rods are one of the most noteworthy materials in the teaching of mathematics throughout the world.

Certainly, Zoltan Dienes, Catherine Stern and George Cuisenaire are not the only mathematics educators who have interpreted number with models. However, they form in a sense a 'Founding Triumvirate' that certainly laid the foundation for what we use today in the teaching and learning of mathematics.

References:

For Zoltan Dienes
https://mathshistory.st-andrews.ac.uk/Biographies/Dienes_Zoltan
The history of base-ten-blocks: Why and who made base-ten-blocks? Rina Kim, Lillie R. Albert, Mediterranean Journal of Social Sciences, vol. 5, no. 9, pp. 356-365, May 2014
For mathematician and teacher Zoltan Dienes, the play was the thing Allison Lawlor, Globe and Mail, Feb. 4, 2014

For Catherine Stern
https://www.encyclopedia.com/women/encyclopedias-almanacs-transcripts-and-maps/stern-catherine-brieger-1894-1973
https://web.archive.org/web/20180406133248/http://www.sternmath.com/who-we-are.html

For George Cuisenaire
https://americanhistory.si.edu/teachingmath/html/302.htm
https://www.froebelweb.org/web2026.htmlboo
http://famousbelgians.net/cuisenaire.htm
file:///C:/Users/fgard/Downloads/ICHME6_DeBock_V3_reluGM+EV.pdf
https://www.eimacs.com/blog/2012/01/georges-papy-mathematics-educator-gifted-math-curriculum/

Levels of Learning: The Impact of Modeling

In using physical models in learning mathematics there are three levels in this process that students need to go through to truly understand mathematics and and the symbolic system by with it is represented.

Concrete Level - What the students do using physical models

At the Concrete Level, the students use only the Base Ten materials to understand mathematics and to solve problems. There is no writing except for writing answers in complete sentences. In this initial step of addressing mathematics with Base Ten materials, the students build an understanding that at the next level will help them realize how to record their work and what the symbolic representations of mathematics really mean.

Transitional Level – How students learn to record the work they are doing with the Base Ten materials using the symbols of mathematics

At the Transitional Level, the students learn how to take the mathematics that they have been modeling with Base Ten materials and communicate what they have done by recording the work using mathematical symbols. In this way, students learn what the symbols as well as their placement really are saying about the mathematics.

Symbolic Level – Students use only mathematical symbols to do the work.

The Symbolic Level is the one that most adults remember learning in school. Many times, we see this as 'the mathematics' and may have forgotten what was done in our learning prior to the development of the symbolism. It is at this level that students are free to express their work in writing, using the Base Ten materials only to verify their ideas if necessary.

It is important for students to move through the Concrete stage of working with Base Ten materials so they can begin to focus on solving problems before the direct use of the symbols.   In a sense, this work gives the students a view of the Base Ten materials to be stored in their 'mind's eye' for use when they will solve problems without the need for Base Ten materials. It is in this way that students learn the symbolic system by which mathematics is communicated.

About This Book

This book focuses on a system by which you as the teacher can assist the students as they transition from the use of Base Ten materials to communicating the work through writing with mathematical symbolism. Here, the focus is working with the two operations with whole numbers, Multiplication and Division.

Each of these operations is treated in their own section so that the teacher can focus on that single topic.   For each operation, the text is separated into two parts.

Part I shows how the Base Ten Materials are used to solve an actual story problem. In this, the problem is solved using the Base Ten materials with diagrams to reflect what the students should be doing with the materials. This is followed by a set of Teacher Practice Problems for you, the teacher, to help you make sure that you understand the process involved before bringing it to your students. And lastly, to help with the implementation of these ideas in your classroom, an outline of a lesson plan for the **Initial Teaching Lesson** is given. As part of the lesson plan, it is suggested that the Teacher Practice Problems be part of an activity during the class for students.

Part II of the writing shows the direct connection between the work with the Base Ten materials and the writing of the algorithm for the operation. In Part II, the same problem from Part I is again used as a model to enhance the connections involved. As each part of the problem is addressed using Base Ten materials, there is an accompanying diagram to show how the work is written using symbols.

Overview of the Book:

The focus of the book is to help you, the teacher, understand the place that physical models (manipulatives) have in mathematics teaching by:

   a. Showing you through problem solving how Base Ten materials can be used by your students to understand mathematics.
   b. Showing you how the models can lead to the symbolic representation of mathematics.
   c. Giving you practice in each of these areas to be ready for your work in your classroom.
   d. Giving you an outline of an **Initial Teaching Lesson** to guide you as you begin to use models with your students.

To review, below is the format for each section which focuses on a whole number operation.

Part I:

   a. **A story problem will be given and solved using only Base Ten materials and a Base 10 Mat on which the students can work.**
   b. **Several Teacher Practice Problems will be given to help you better understand the ideas of solving problems using Base Ten materials.**
   c. **For use in your classroom teaching, an outline of an Initial Teaching Lesson will give you the information as to how to implement problem solving with Base Ten Materials in your classroom.**

Part II:

   d. **The same story problem used in Part I will again be given and solved. Here the link between the use of Base 10 materials and the writing (recording) of what occurs will be shown.**
   e. **The same Teacher Practice Problems from Part I will be given for you to better connect the work of solving problems using Base Ten materials to writing the symbolism and algorithm for that operation.**
   f. **For use in your classroom teaching, an outline of an Initial Teaching Lesson will give you the information as to how to implement the connection between the use of Base Ten Materials to the written symbolism and algorithm for that operation. To assist with this, a Recording Sheet is included for use by the teacher in the modeling and by students in later work.**

Teaching Multiplication of a 2-Digit by a 1-Digit Whole Number Using Base-10 Materials

Multiplication: Trading Ones for Tens

PART I CONCEPTUAL DEVELOPMENT

In addressing how to solve multiplication problems with models the first idea is to demonstrate how the students develop the concepts that underpin the multiplication of a 2-digit number by a 1-digit number with regrouping, or as we call it here, trading. The activity here allows the students to physically work with the quantities and see the rationale for the need for trading and what happens to the ten that they trade for.

As with any problem solving, it begins with a story. This story will help you understand the concepts involved as well as form the basis for the Initial Teaching Lesson that is provided at the end of Part I.

The Story

In a basketball tournament, Allie played in 3 games and scored 24 points in each game. Her younger sister, Vicki, wanted to find out the total number of points Allie scored. Vicki decided to make a model and find the answer.

Modeling the Story.

To model the story, Vicki used Base 10 Materials.

Since Allie scored 24 points three times, Vicki showed 24 using 2-tens and 4-ones **three times** on her Base-10 Mat.

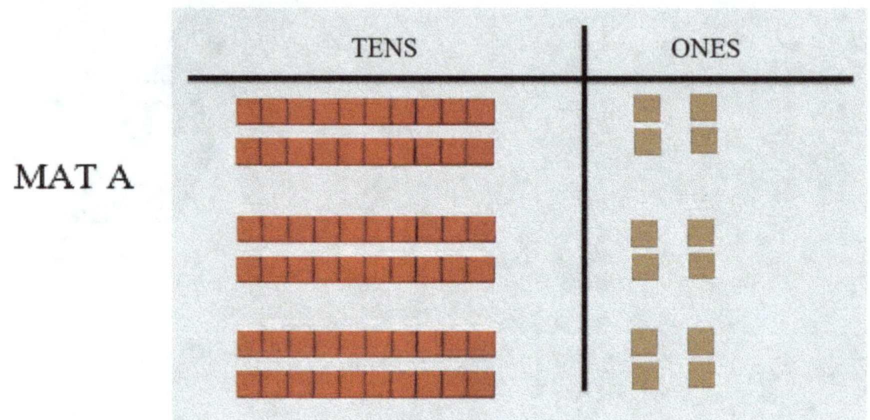

Putting The Ones Together

Vicki now put all the ones together.

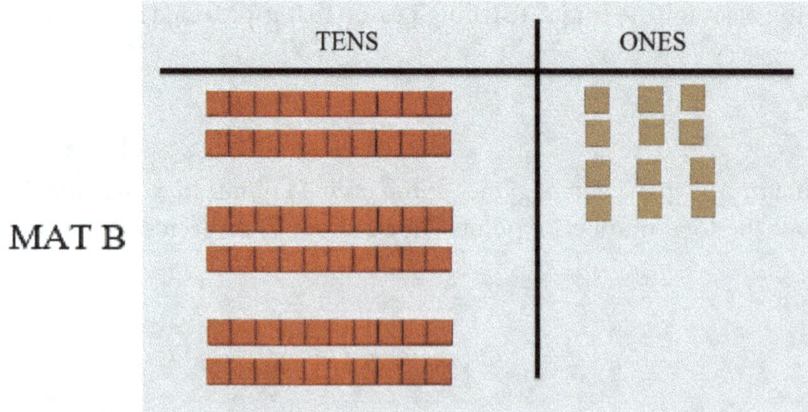

When she counted them, she found that she had 12-ones.

MULTIPLICATION AND DIVISION OF 2-DIGIT NUMBERS BY 1-DIGIT NUMBERS

Making a New Ten

To have less pieces on the mat, Vicki decided to trade 10-ones for a new ten.

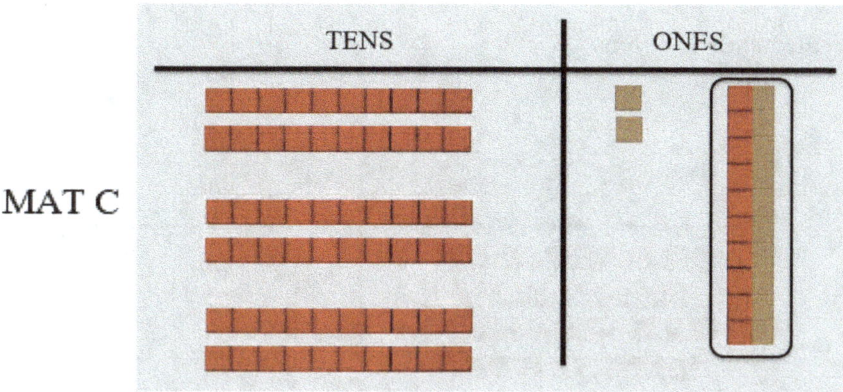

She then took away the 10-ones and put the new ten above the other tens in the tens column.

This left her with 2-ones.

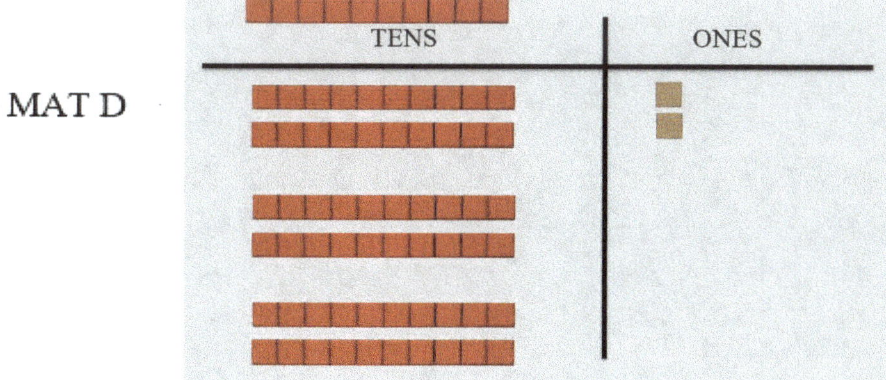

Putting the Tens Together

Vicki now put the tens together.

She had 3 sets of 2-tens. That gave her 6-tens. Then she added in the new ten.

When she put them together, she had 7-tens

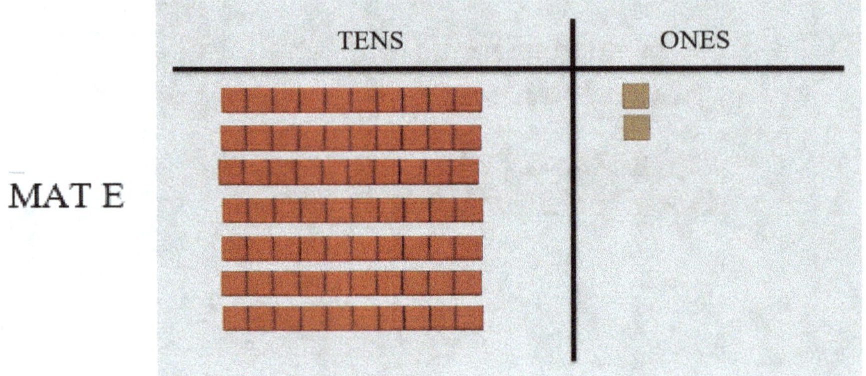

So Vicki now had 7-tens and 2-ones which equals 72.

So, Allie scored 72 points in the tournament.

Teacher's Practice: Here are some problems for you to try using only the base ten materials. The reason for these exercises is to have you assess your knowledge of their use and your ability to use them to solve problems. For these, concentrate on each of the MATS that you produce and the reasoning behind them. This will help in the next section when you review a sample Initial Teaching Lesson Plan that you may wish to use in your classroom. Also, to be a good model for your students, write the answer to each problem in a complete sentence.

1. In helping deliver school materials to the classrooms, Eli placed 26 notebooks in 3 of the fourth-grade rooms at Meadows Elementary School. What was the total number of notebooks delivered by Eli?
2. Maria filled 2 boxes with basketball uniforms. Each box contained 18 uniforms. How many uniforms did Maria put in the boxes?
3. As league president, Hector wanted to give each team 16 baseballs for use during practice. If there are 6 teams in the league, how many baseballs does Hector need to buy?
4. For a field trip to the museum, Ms. Sanchez ordered 3 buses. Each bus will carry 24 people. How many people can go on the fieldtrip?
5. To help deliver flyers to announce that he was running for school president, Zach gave each of his 3 friends 29 flyers. How many flyers did Zach give to his friends?

Outline of an 'Initial Teaching Lesson'

Objective: To have students work with Base Ten materials to solve a problem involving the multiplication of a 2-digit number by a 1-digit number using trading.

Materials: For each student, provide a set of 9 ten-bars and 19 ones.

A Place Value Mat. (A Model is given on the page after the lesson outline.)

Lesson Outline:

I Read or display the first sentence of the story.

'In a basketball tournament, Allie played in 3 games and scored 24 points in each game.'

Ask the students to show '24 point' on their Mats three times. (See **MAT A** above.)

After they do this, discuss why they should have 3 groups of 2-tens and 4-ones on their Mat.

II Now read or display the second sentence of the story.

'Her younger sister, Vicki, wanted to find out the total number of points Allie scored.'

To begin, have the students put all the ones together (**Mat B**).

III Have them count the ones and ask them it they need to trade for a new ten.

Since there are 12 ones, a trade is needed.

IV Have the students match up 10 of the ones for a new ten to begin trading process. **(Mat C)**

V Have the students trade the 10 ones for the new ten. Have them remove the 10 ones and place the new ten above the mat next to the word, TENS. **(Mat D)**

VI Now have the students count the number of tens that they had on the mat and then add in the new ten. **(Mat E)**

VII Have the students discuss the value of the tens and ones on their mats (62) and how this solves the problem.

VIII As a group, discuss the work and review what the students did with the materials **(Mat A to Mat E)**

IX Have the students write the answer in a complete sentence.

> **'Allie scored a total of 72 points in the three games.'**

X The students can be given the problems from the Teacher's Practice section above or from the textbook/program that is in use in the classroom. During this, the students will complete the work using the Base Ten materials and write each answer in a complete sentence. For some of the problems, the students can work in small groups while for other problems you may wish to have them complete the work individually as a type of formative assessment.

ONES

TENS

USING SYMBOLS TO RECORD THE WORK

SOME HISTORY ABOUT MULTIPLICATION SYMBOLS

The Symbol for Multiplication

The multiplication symbol, 'x', dates back to the 1600's used by mathematicians John Napier and William Oughtred.

There are other 'signs' that are used for multiplication as we move through mathematics.

One of the most used is parentheses where the expression, (3)(4) means 3 times 4.

In algebra, if there is only one number involved, then just writing the symbols side-by-side indicates multiplication.

For example, 3b means the value of 3 times the value of b.

Also, if you have two variables, 'c' and 'd,' then 'cd' means you multiply the value of 'c' times the value of 'd.'

For obvious reasons, this symbolism is not used with numbers since it has been established that 34 means thirty-four and not 3 times 4.

So, with this information, let's begin to use Base 10 Materials to solve problems and then move on to record the work remembering that mathematical symbolism was and continues to be created by people.

PART II CONCEPTUAL AND SKILL DEVELOPMENT

Recording Our Work to Communicate Mathematics

Part II demonstrates the link between the work using the Base Ten materials and how this work can be recorded in ways that will allow the mathematics to be communicated to others.   The writing here shows how the physical use of the Base Ten materials is recorded on paper using the symbolism by which mathematics is communicated.

As with any problem solving, it begins with a story. The work solving the story will help you understand how the work with Base Ten materials is linked to the algorithm for multiplication of a 2-digit number by a 1-digit number. This will also form the basis for the Initial Teaching Lesson that is provided at the end of Part II.

The Story

In a basketball tournament, Allie played in 3 basketball games and scored 24 points in each game. Her younger sister, Vicki, wanted to find out the total number of points Allie scored. Vicki decided to make a model and find the answer.

Modeling the Story.

To model the story, Vicki used base ten materials. She also recorded her work on a tens/one chart.

Vicki made this model on a Base-10 Mat using her materials.

Since Allie scored 24 points three times, Vicki showed
24 using 2-tens and 4-ones **three times** on her tens/ones mat.

Since she was multiplying, she recorded:

MAT A

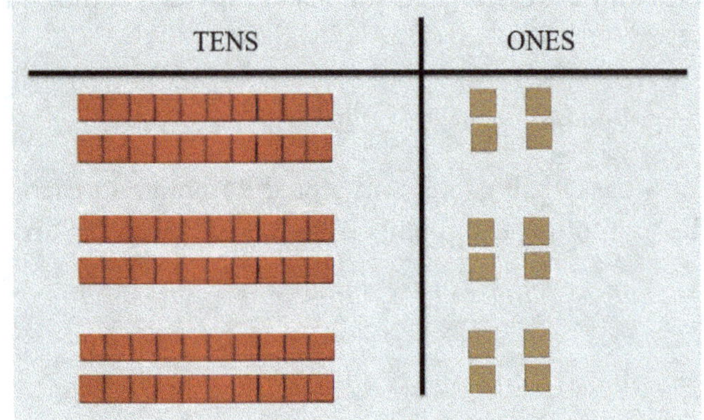

LINK A

TENS	ONES
2	4
x	3

MULTIPLICATION AND DIVISION OF 2-DIGIT NUMBERS BY 1-DIGIT NUMBERS

Putting The Ones Together

Vicki now put all the ones together.

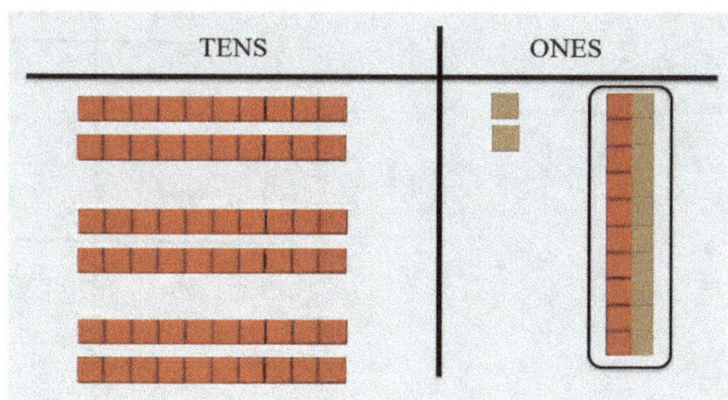

MAT B

LINK B

TENS	ONES
2	4
x	3

When she counted them, she found that she had 12-ones.

Making a New Ten

To have less pieces on the mat, Vicki decided to trade 10-ones for a new ten.

MAT C

LINK C

TENS	ONES
2	4
x	3

So she took away the 10-ones and put the new ten above the other tens in the tens-column. This left her with 2-ones.

She then wrote '2' in the one-column under the line and wrote a '1' at the top of the tens-column to show that she had made a new ten.

MAT D

LINK D

TENS	ONES
1	
2	4
x	3
	2

Putting the Tens Together

Vicki now put the tens together.

She had 3 sets of 2-tens. That gave her 6-tens.
Then she added in the new ten.
When she put them together, she had 7-tens

She wrote a '7' in the tens-column under the line.

MAT E

LINK E

TENS	ONES
1	
2	4
x	3
7	2

So Vicki now had 7-tens and 2-ones on the tens-ones mat and 72 on the tens/ones chart.

Allie scored 72 points in the tournament.

Teacher's Practice: Here are some problems for you to try using base ten materials and relating them to the writing of the algorithm. The reason for these exercises is to have you assess your knowledge of their use and your ability to use them to solve problems. For these, concentrate on each of the MATS that you produce, the reasoning behind them and their linkage to the writing of the algorithm for multiplication of a 2-digit number by a 1-digit number. This will help in the next section when you review a sample Initial Teaching Lesson Plan that you may wish to use in your classroom to help your students establish this linkage. Again, to be a good model for your students, write the answer to each problem in a complete sentence.

(Please note: These problems are the same as those from Part I to allow you to focus on the important linkage of the Base Ten materials and the written algorithm.)

1. In helping deliver school materials to the classrooms, Eli placed 26 notebooks in 3 of the fourth-grade rooms at Meadows Elementary School. What was the total number of notebooks delivered by Eli?
2. Maria filled 2 boxes with basketball uniforms. Each box contained 18 uniforms. How many uniforms did Maria put in the boxes?
3. As league president, Hector wanted to give each team 16 baseballs for use during practice. If there are 6 teams in the league, how many baseballs does Hector need to buy?
4. For a field trip to the museum, Ms. Sanchez ordered 3 buses. Each bus will carry 24 people. How many people can go on the fieldtrip?
5. To help deliver flyers to announce that he was running for school president, Zach gave each of his 3 friends 29 flyers. How many flyers did Zach give to his friends?

Outline of an 'Initial Teaching Lesson'

Objective: To have students understand the link between the work they did with the Base Ten materials with multiplication of a 2-digit number by a 1-digit number and how to record this work on paper.

Materials:

For each student:

a. **Provide a set of 9 ten-bars and 19 ones.**
b. **A Place Value Mat. (Students can use the Mat from their previous work.)**
c. **A Tens-Ones Recording Page for students to use in recording the symbols as they are developed in solving the problems.**

For the teacher:

A Tens-Ones Recording Page for your use in recording the symbols as they are developed. (A Model is given on the page after the lesson outline.)

<u>NOTE</u>: **In this lesson, the teacher acts as the recorder while the students solve the problem the problem using Base Ten materials.**

Lesson Outline:

I Read or display the first sentence of the story.

'In a basketball tournament, Allie played in 3 basketball games and scored 24 points in each game.'

Ask the students to show '24 point' on their Mats three times. (See **MAT A** above.)

After they do this, discuss why they should have 3 groups of 2-tens and 4-ones on their Mat.

On the **Tens-Ones Recording Page,** record 24 x 3 in vertical form. **(LINK A)**

II Now read or display the second sentence of the story.

'Her younger sister, Vicki, wanted to find out the total number of points Allie scored.'

To begin, have the students put all the ones together **(MAT B) (For this, there is no writing on the Tens-Ones Recording Page).**

III Have them count the ones and ask them it they need to trade for a new ten.

Since there are 12 ones, a trade is needed. **(For this, there is no writing on the Tens-Ones Recording Page).**

IV Have the students match up 10 of the ones for a new ten to begin trading process. **(Mat C)**

V Have the students trade the 10 ones for the new ten. Have them remove the 10 ones and place the new ten above the mat next to the word, TENS. **(MAT D)** Record this by writng a '2' below the line in the ONES column and writing a '1' above the '2' 'TENS' to represent the new ten. **(LINK D)**

VI Now have the students count the number of tens that they had on the mat and then add in the new ten. **(MAT E)** Record this by writing a '7' below the line in the TENS column. **(LINK E)**

VII Have the students discuss the value of the tens and ones on their mats (72) and how this solves the problem.

VIII As a group, discuss the work and review what the students did with the materials reviewing how you recorded the work **(Mat D to Mat E.)**

IX Have the students write the answer in a complete sentence.

    'Allie scored a total of 72 points in the three games.'

X The students can be given the problems from the Teacher's Practice section above or from the textbook/ program that is in use in the classroom. Initially have the students work in pairs. For each problem one of the students will use the Base Ten Materials to show the work and the other student will use the **Tens-Ones Recording Page** to record the work. For each problem, the students will agree on a complete sentence to answer the question.

Tens/Ones Recording Page For The Multiplication Algorithm

TENS	ONES

Teaching Division of a 2-Digit by a 1-Digit Whole Number Using Base-10 Materials

Division: Trading Tens for Ones

PART I CONCEPTUAL DEVELOPMENT

In addressing how to solve division problems with models the first idea is to demonstrate how the students develop the concepts that underpin the division of a 2-digit number by a 1-digit number with regrouping, or as we call it here, trading. The work here allows the students to physically work with the quantities and see the rationale for the need for trading and what happens to the ones that they trade for.

As with any problem solving, it begins with a story. This story will help you understand the concepts involved as well as form the basis for the Initial Teaching Lesson that is provided at the end of Part I.

The Story

In helping the school librarian, Jackie is asked to share 45 books with the fourth-grade classrooms. There are 4 boxes that contain 10 books in each box and there are 5 single books.

There are 3 fourth grade classrooms, Room 104, Room 105 and Room 106.

If each classroom is to receive the same number of books, Jackie has to decide how many books she will give to each class.

Modeling the Story.

So, to model the story, Jackie used base ten materials. She used 4-tens to model each box of ten books and 5-ones to model the 5 single books.

Here is a model that Jackie developed using her materials on a Base-10 Mat.

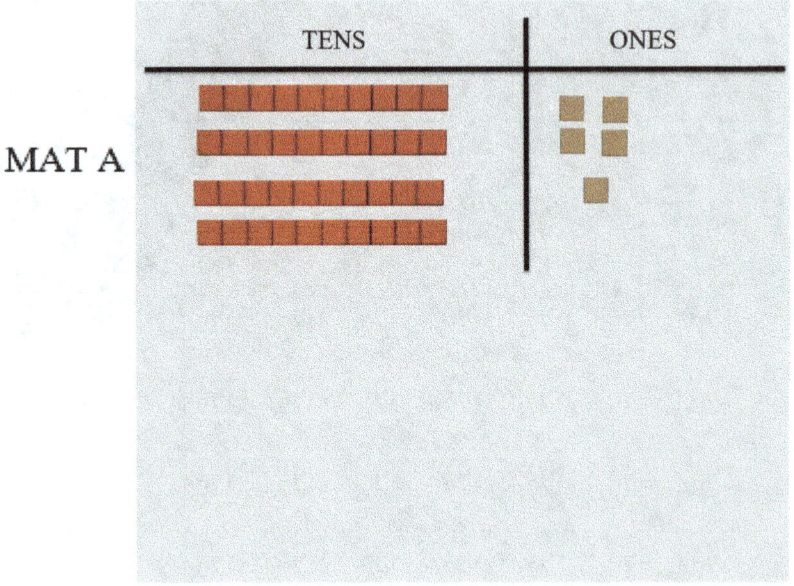

Since there were three classrooms, Jackie used a card to represent each classroom and placed the cards at the bottom of the Base-10 Mat.

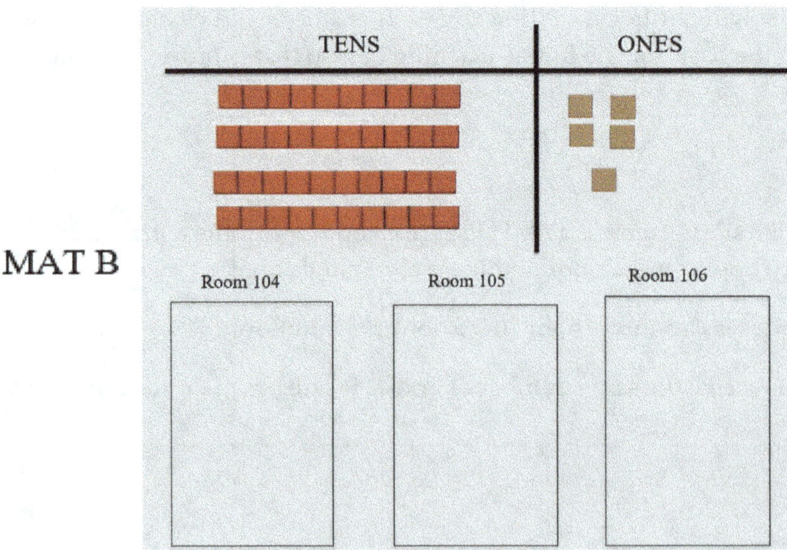

Sharing the Tens

Now Jackie saw that she could share 1 box of 10 books with each classroom.

So she showed this using the model by putting 1-ten (1 box of books) on each card.

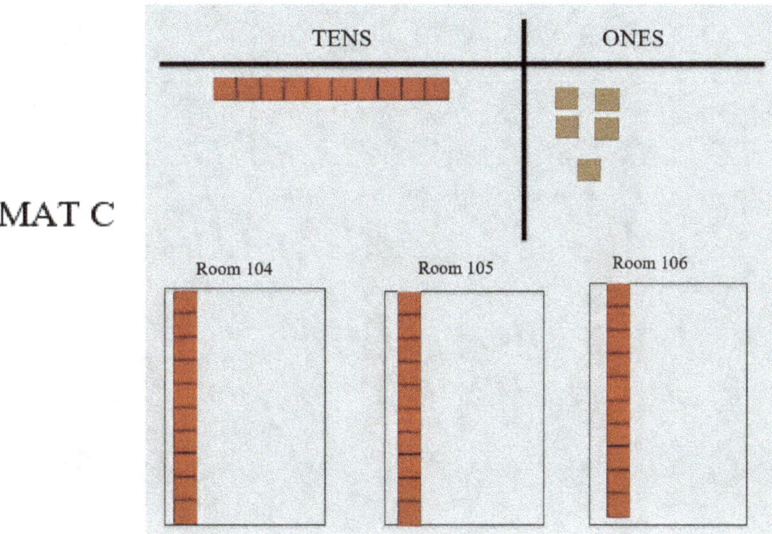

Sharing the Last Ten

But Jackie knew that she could not share the last box as a full box because then one room would receive more books than the others.

So she knew that she would have to open the last box and take the 10 books out.

To show this using the model, Jackie traded the remaining ten-bar for 10-ones.

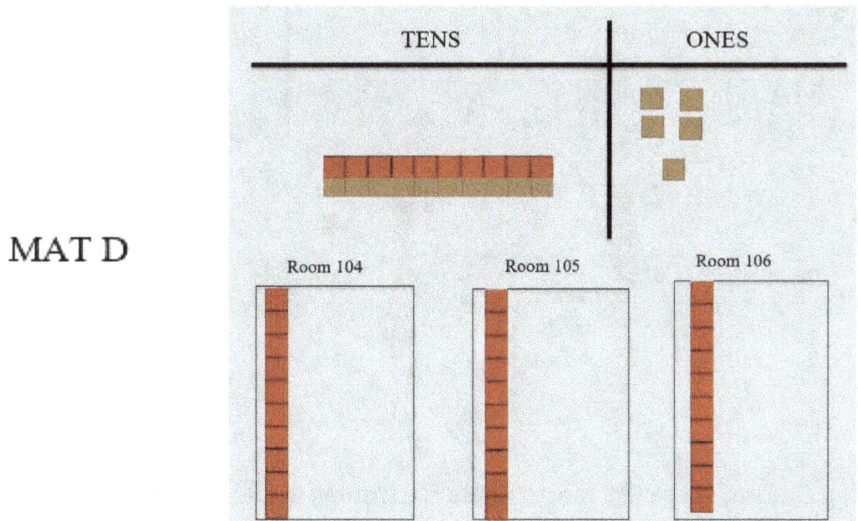

She then placed the new ones in the ones-column on the Base-10 Mat

Here is what she had on her Base-10 Mat

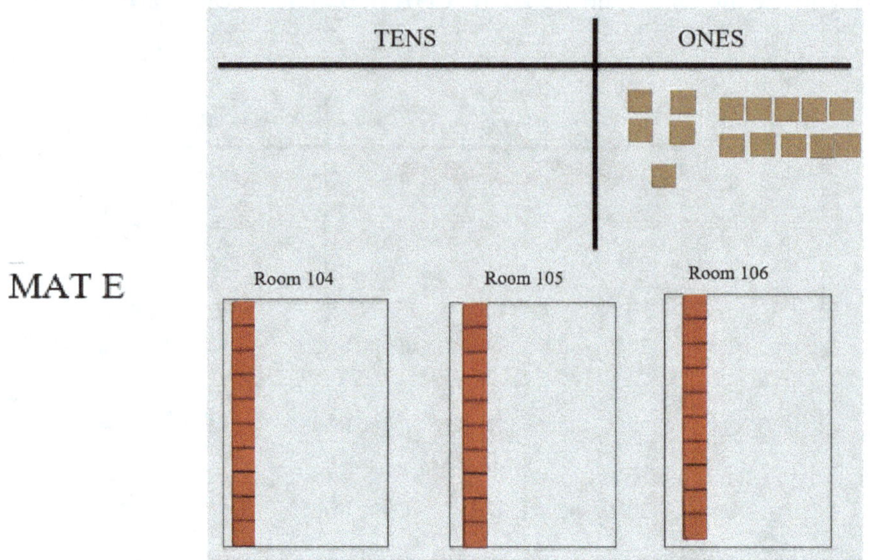

Sharing the Ones

Now Jackie saw that she had 15-ones which represented 15 books. She knew that she could share these 15 books (ones) with the 3 classrooms.

So in her model, she shared out the ones equally.

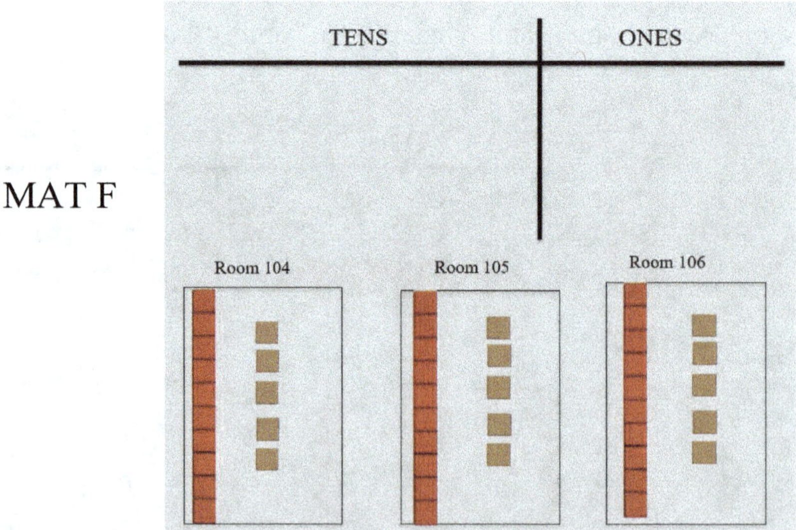

From her model, Jackie saw that the 3 fourth grade classrooms would each receive 15 books.

  MULTIPLICATION AND DIVISION OF 2-DIGIT NUMBERS BY 1-DIGIT NUMBERS

Teacher's Practice: Here are some problems for you to try using only the base ten materials. The reason for these exercises is to have you assess your knowledge of their use and your ability to use them to solve problems. For these, concentrate on each of the MATS that you produce and the reasoning behind them. This will help in the next section when you review a sample Initial Teaching Lesson Plan that you may wish to use in your classroom. Lastly, to be a good model for your students, write the answer to each problem in a complete sentence.

1. Mariel is going to fill 3 boxes with apples for the food pantry in the community. Each box is to have the same number of apples. When Mr. Alvarez delivered the apples to her, they were 48 apples. There were 4 packs of 10 apples and 8 other apples. How many apples did Mariel put in each box?
2. Jess wraps one fork and one knife in a napkin to place on the tables in the restaurant where he works. He makes 56 of these sets and will place 4 sets on each table. How many tables can he set up with forks and knives?
3. In accepting a delivery, Gerry has 96 bottles of juice. He receives 9 boxes containing ten bottles each and another box with 6 bottles of juice. For their hike, he wants to give each hiker 4 bottles of juice. To how many people can he supply juice?
4. As the league president, Mazie bought 90 baseballs. She will give them out equally to each of the 6 teams in the league. How many baseballs will each team receive?

Outline of an 'Initial Teaching Lesson'

Objective: To have students work with Base Ten materials to solve a problem involving the division of 2-digit number by a 1-digit number using trading.

Materials: For each student, provide a set of 9 ten-bars and 19 ones.

> **A Place Value Mat. (A Model is given on the page after the lesson outline.)**

> **3 cards to represent the 3 classrooms.**

Lesson Outline:

I Read or display the first two sentences of the story.

"In helping the school librarian, Jackie is asked to share 45 books equally with the fourth- grade classrooms. There are 4 boxes that contain 10 books in each box and there are 5 single books."

Ask the students to show '45' on their Mats. (See **MAT A** above.)

After they do this, discuss why they should have put 4-tens and 5-ones on their Mat.

II Now read or display the third sentence of the story.

There are 3 fourth grade classrooms, Room 104, Room 105 and Room 106.

Have the students represent each of these classrooms by placing the three cards at the bottom of the Mat, one for each classroom. You may have the children assign 'classroom numbers' to the cards. (**Mat B**)

III   Now read or display the fourth sentence of the story.

"If each classroom is to receive the same number of books, Jackie has to decide how many books she will give to each class."

Have the students work in small groups (3 students) or individually using the Base Ten materials to find how many books Jackie will put in each classroom.

Note: Remind the students that it is okay if they need to trade a ten-bar for 10 ones.

IV   Give the students time to work on this. Then discuss what the groups did to solve the problem. At their places, ask them to write the answer in a complete sentence.

V   Have a student use either Base Ten materials and a document camera or a set of Base Ten materials for the chalkboard to explain what the student did.

VI   Discuss the student's work and have others show their methods.

VII   As a review of the process, demonstrate how you would use the Base 10 materials to solve the problem using the explanations and **MAT C** through **MAT F** above. (In doing this, it will give the students a basis for the later work of linking the use of the Base Ten materials to the writing of the algorithm.) You may wish to have the students follow along with you and use their materials as you use yours.

VIII   The students can be given the problems from the Teacher's Practice section above or from the textbook/program that is in use in the classroom. During this, the students will complete the work using the Base Ten materials and write each answer in a complete sentence. For some of the problems, the students can work in small groups while for other problems you may wish to have them complete the work individually as a type of formative assessment.

ONES

TENS

USING SYMBOLS TO RECORD THE WORK

SOME HISTORY ABOUT DIVISION SYMBOLS

The Symbol for Division

While we can solve division 'stories' using Base-10 materials, we need to have a system of symbols to record and communicate the results. For division, we use a symbol that we call a 'division bracket.' The quantities in a scenario that are used to solve a problem with division are placed in certain positions in and around the division bracket.

Using Jackie's work with the books, we have three quantities of which we know two of them. We know the **Total** number of books (45) and the number of classrooms (3) that will receive the books. (We could call these classrooms the **Sharers**.) The unknown quantity is the number of books that each classroom will receive. The division bracket is drawn using two symbols, a right parenthesis, ')', and a line above the Total that is called a 'viniculum' which is a symbol that shows that these numerals are all together. Here is what the division Bracket looks like

and here is where quantities are placed:

Amount each receives

Sharers)   Total

Now, for Jackie to record the work done on the Base-10 Mat and the cards, we begin with the division bracket:

$$3 \overline{)\ 4\ 5}$$

To show how the division Bracket is used, we place it on a Base-10 Symbols Chart

Tens | Ones

$$3 \overline{)\ 4\,|\,5}$$

So in our work here, we will link the work with Base 10 materials using the Division Bracket as a recording device.

PART II CONCEPTUAL AND SKILL DEVELOPMENT

Recording Our Work to Communicate Mathematics

Part II demonstrates the link between the work using the Base Ten materials and how this work can be recorded in ways that will allow mathematics to be communicated to others.   The writing here shows how the physical use of the Base Ten materials is recorded on paper using the symbolism by which mathematics is communicated.

As with any problem solving, it begins with a story. The work solving the story will help you understand how the work with Base Ten materials is linked to the algorithm for division of a 2-digit number by a 1-digit number. This will also form the basis for the Initial Teaching Lesson that is provided at the end of Part II.

The Story

In helping the school librarian, Jackie is asked to share 45 books with the fourth-grade classrooms. There are 4 boxes that contain 10 books in each box and there are 5 single books.

There are 3 fourth grade classrooms, Room 104, Room 105 and Room 106.

If each classroom is to receive the same number of books, Jackie has to decide how many books she will give to each class.

Modeling the Story.

So, to model the story, Jackie used base ten materials. She used a ten-bar to model each box of books and 5-ones were used to model the 5 single books.

Here is a model that Jackie developed using her materials on a Base-10 Mat.

On the Recording Chart using the 'division bracket' she writes '4 in the tens column' and '5 in the ones column' to show the total number of books which is 45.

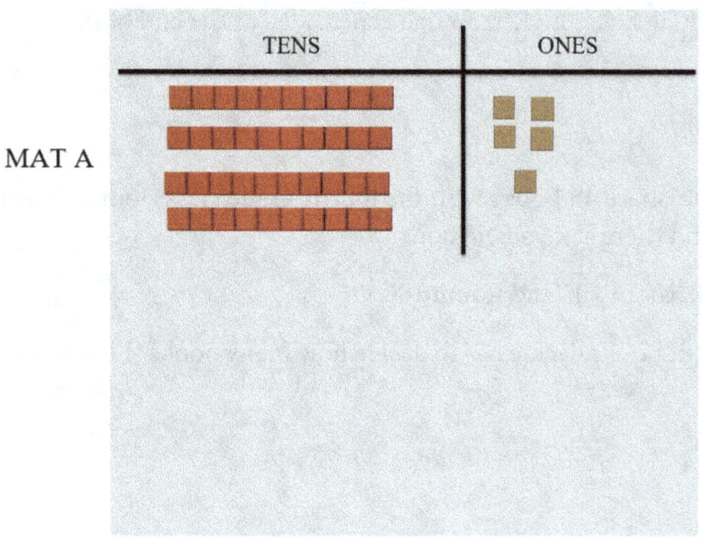

Since there were three classrooms, Jackie used a card to represent each classroom and placed the cards at the bottom of the Base-10 Mat.

She wrote a '3' outside of the division bracket on the Recording Chart to show the 3 classrooms.

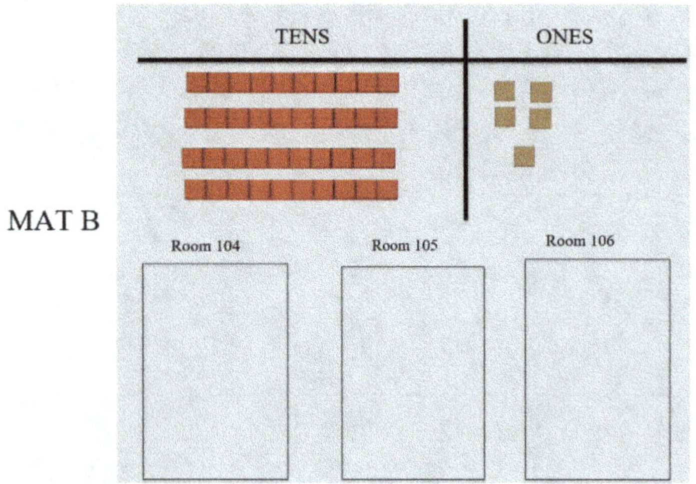

Sharing the Tens

Now Jackie saw that she could share
1 box of 10 books with each classroom.
So she showed this using the model by
putting 1-ten on each card.

She recorded this by writing a '1'
in the Tens' Column above
the line on the Recording Mat.

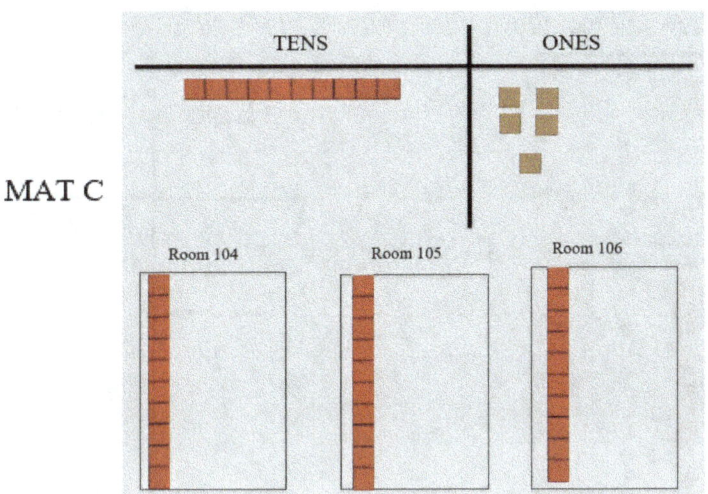

LINK C

Then on the Recording Mat, she recorded
how many tens she shared (3) and showed
the subtraction of these 3-tens from the
total of 4-tens. This now showed that she
had 1-ten (1 box of books) left.

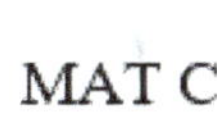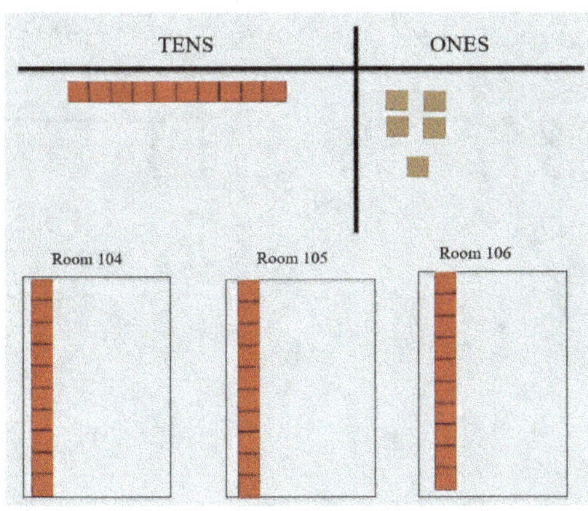

LINK C-1

Sharing the Last Ten

But Jackie knew that she could not share the one remaining full box because then one room would receive more books than the others.

She knew that she would have to open the last box and take the 10 books out.

To show this using the model, Jackie traded the last ten-bar for 10-ones.

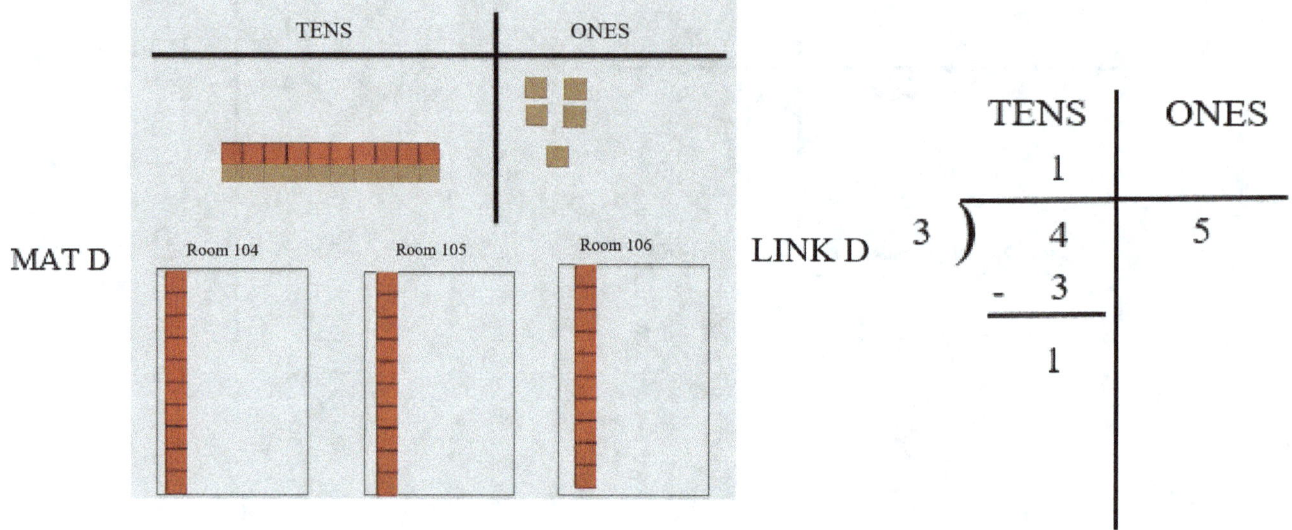

This gave her 15 ones in the ones column.

To record this on the Chart, she drew an arrow to show that the 5 books would be combined with the 10 books for a total of 15 books.

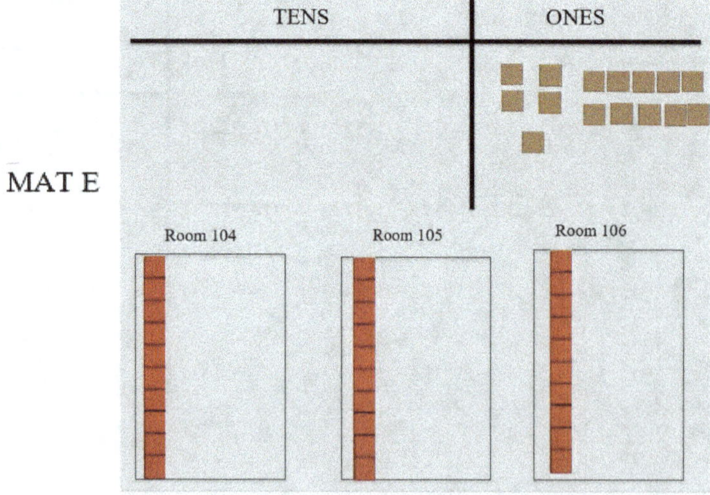

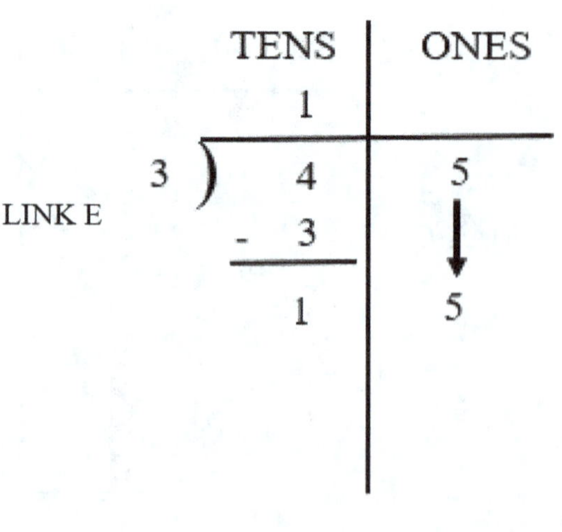

MULTIPLICATION AND DIVISION OF 2-DIGIT NUMBERS BY 1-DIGIT NUMBERS

Sharing the Ones

Now Jackie had 15-ones. She knew that if she shared them equally with the 3 classrooms, each classroom would receive 5 books.

So in her model, she shared the ones.

She recorded this by writing a '5' in the ones-column above the line on the Recording Chart

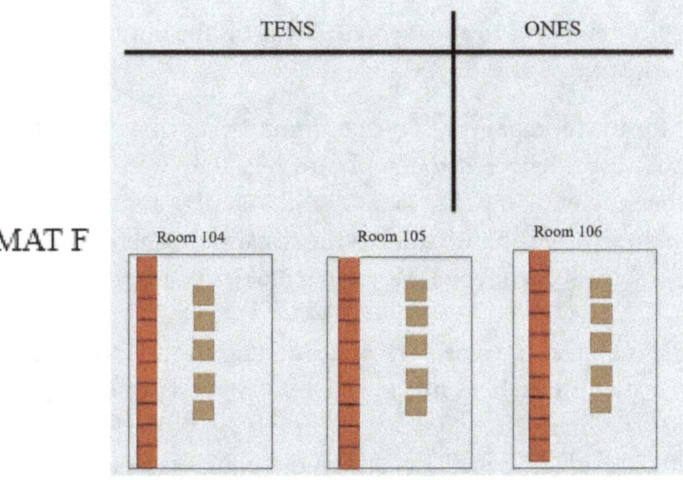

On her Recording Chart, she then subtracted the 15-ones that were shared with the 3 classes. This showed that there were no books remaining.

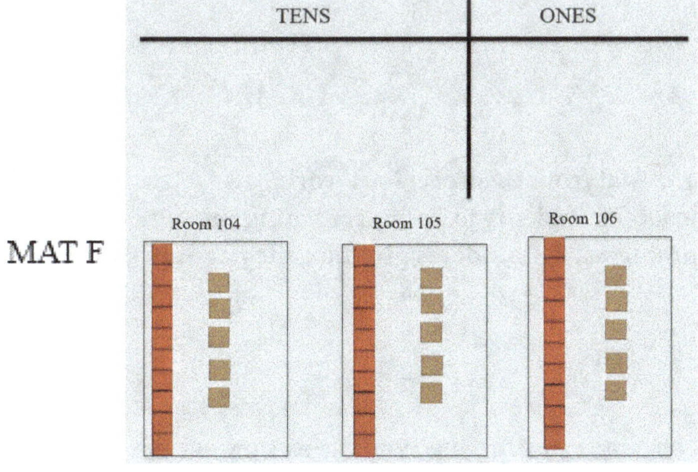

From her model, Jackie saw that the 3 fourth grade classrooms would each receive 15 books.

The number of books that each classroom would receive is seen above the line on the Recording Chart. She has now recorded her work using symbols. So she can now communicate what she did.

Teacher's Practice: Here are some problems for you to try using base ten materials and relating them to the writing of the algorithm. The reason for these exercises is to have you assess your knowledge of their use and your ability to use them to solve problems. For these, concentrate on each of the MATS that you produce the reasoning behind them and their linkage to the writing of the algorithm for dividing a 2-digit number by a 1-digit number. This will help in the next section when you review a sample Initial Teaching Lesson Plan that you may wish to use in your classroom to help your students establish this linkage. Again, to be a good model for your students, write the answer to each problem in a complete sentence.

(Please note: These problems are the same as those from Part I to allow you to focus on the important linkage of the Base Ten materials and the written algorithm.)

1. Mariel is going to fill 3 boxes with apples for the food pantry in the community. Each box is to have the same number of apples. When Mr. Alvarez delivered the apples to her, they were 48 apples. There were 4 packs of 10 apples and 8 other apples. How many apples did Mariel put in each box?
2. Jess wraps one fork and one knife in a napkin to place on the tables in the restaurant where he works. He makes 56 of these sets and will place 4 sets on each table. How many tables can he set up with forks and knives?
3. In accepting a delivery, Gerry has 96 bottles of juice. He receives 9 boxes containing ten bottles each and another box with 6 bottles of juice. For their hike, he wants to give each hiker 4 bottles of juice. To how many people can he supply juice?
4. As the league president, Mazie bought 90 baseballs. She will give them out equally to each of the 6 teams in the league. How many baseballs will each team receive?

Outline of an 'Initial Teaching Lesson'

Objective: To have students understand the link between the work they did with the Base Ten materials with Division of 2-digit number by a 1-digit number and how to record this work on paper.

Materials:

For each student provide:

a. **A set of 9 ten-bars and 19 ones.**
b. **A Place Value Mat. (Students can use the Mat from their previous work.)**
c. **A Recording Page with a Division Bracket for students to use in recording the symbols as they are developed in solving the practice problems. (A Model is given on the page after the lesson outline.)**

For the teacher:

A Recording Page with a Division Bracket for use in recording the symbols as they are developed. (A Model is given on the page after the lesson outline.)

NOTE: In this lesson, the teacher acts as the recorder while the students solve the problem using Base Ten materials. Later, students can use the Recording Page with a Division Bracket as they practice with other problems.

Lesson Outline:

I   Read or display the first two sentences of the story.

"In helping the school librarian, Jackie is asked to share 45 books equally with the fourth- grade classrooms. There are 4 boxes that contain 10 books in each box and there are 5 single books."

Ask the students to show '45' on their Mats. After they do this, discuss why they should have put 4-tens and 5-ones on their Mat.

On the **Recording Page with a Division Bracket,** write a '4' in the TENS column in the Division Bracket and a '5' in the ONES column in the Division Bracket. (**Link A**)

II   Now read or display the third sentence of the story.

There are 3 fourth grade classrooms, Room 104, Room 105 and Room 106.

Have the students represent each of these classrooms by placing cards at the bottom of the Mat. On the **Recording Page with a Division Bracket,** write a '3' on the outside of the Division Bracket. (**Link B**)

III   Now read or display the fourth sentence of the story.

"If each classroom is to receive the same number of books, Jackie has to decide how many books she will give to each class."

Have the students discuss how they completed this work previously.

IV   Now, have the students set up the problem again. As they move through each step, show them how it can be recorded.

They first shared the Tens. Each class (card) received 1-ten. Show how to record this by writing a '1' in the TENS column above the Division Bracket. (**Link C**)

V   To show the number of Tens that remain, write '-3' under the '4' in the TENS column in the **Recording Page with a Division Bracket**. Complete the subtraction to show a '1' in the TENS column at the bottom of the Division Bracket. (**Link C-1**)

VI Since there are not enough tens to share, have the students trade the remaining ten for 10-ones and place them with the other ones in the ONES column of the Division Bracket.

On the **Recording Page with a Division Bracket,** show this by showing the '5' in the ONES column moving down to be placed next to the '1' in the TENS column. (**LINK E**)

VII   The students now share the ones so that each card receives 5 ones.

To show this on the **Recording Page with a Division Bracket,** write a '5' in the ONES column above the Division Bracket. **(LINK F)**

VIII   Since the students have now shared all of the ones, show this by writing '-15' at the bottom of the Division Bracket under the original '15.' This shows that when you subtract, there are no ones remaining. **(LINK F-1)**

IX   Review the student's work and the recording method with the class.

VII The students can be given the problems from the Teacher's Practice section above or from the textbook/program that is in use in the classroom. During this, the students will complete the work using the Base Ten materials and write each answer in a complete sentence. For some of the problems, the students can work in small groups while for others you may wish to have them complete the work individually as a type of formative assessment. Have the students alternate from working with the materials to being the recorder so they can discuss the relationship between these.

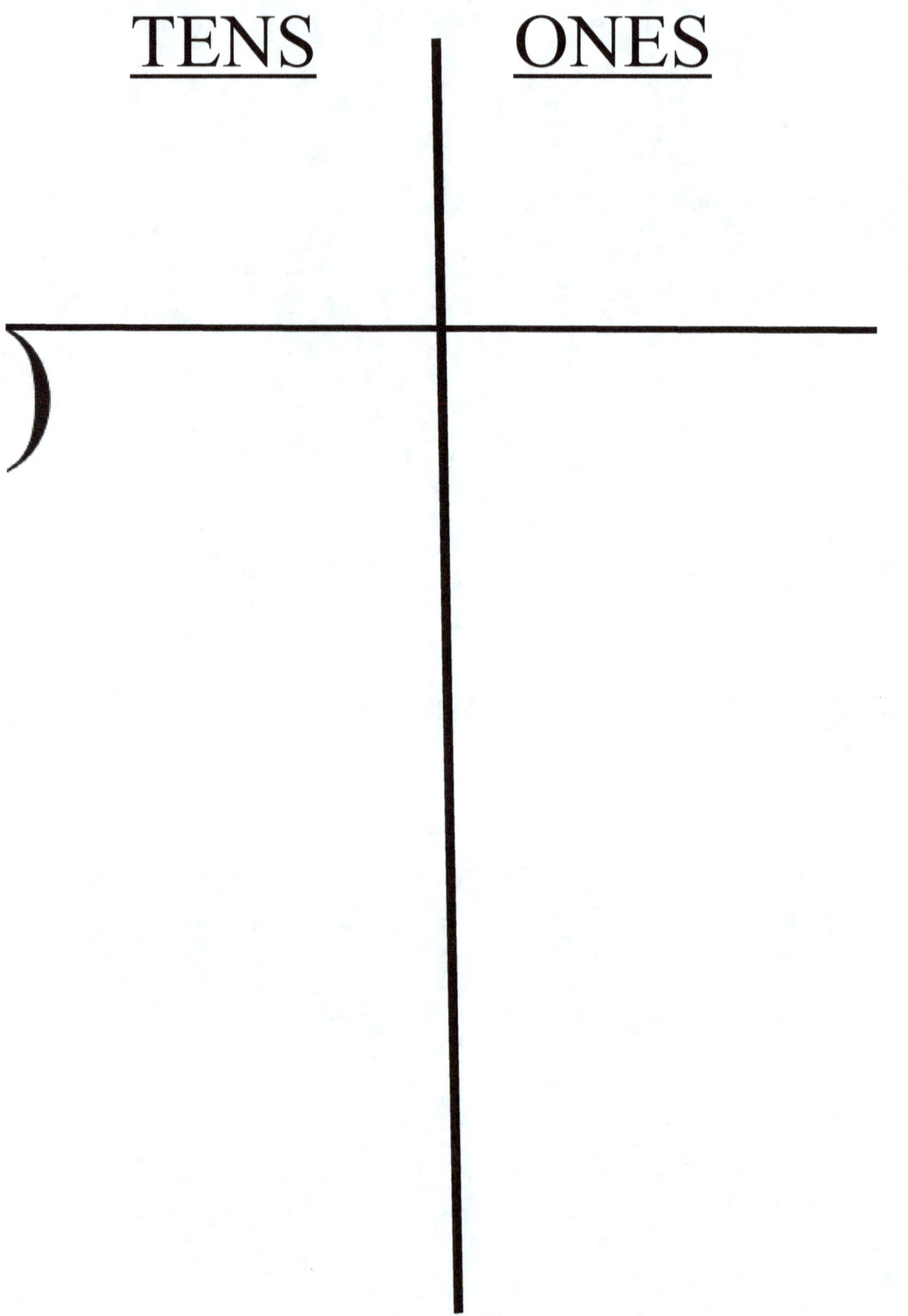
TENS
ONES